AF269754

XTREME SURVIVAL

SURVIVING THE SWAMP

A&D Xtreme
BOLD HI-LO NONFICTION

An imprint of Abdo Publishing
abdobooks.com

JESSICA RUSICK

TAKE IT TO THE XTREME!

GET READY FOR AN EXTREME ADVENTURE!
THE PAGES OF THIS BOOK WILL TAKE YOU INTO
THE THRILLING WORLD OF OUTDOOR SURVIVAL.
WHEN YOU HAVE FINISHED READING THIS BOOK, TAKE THE
XTREME CHALLENGE ON PAGE 45 ABOUT WHAT YOU'VE LEARNED!

ABDOBOOKS.COM

Published by Abdo Publishing, a division of ABDO, PO Box 398166, Minneapolis, Minnesota 55439.
Copyright ©2024 by Abdo Consulting Group, Inc. International copyrights reserved in all countries.
No part of this book may be reproduced in any form without written permission from the publisher.
A&D Xtreme™ is a trademark and logo of Abdo Publishing.
Printed in the United States of America, North Mankato, MN.
102023
012024

THIS BOOK CONTAINS
RECYCLED MATERIALS

Design: Tamara JM Peterson, Mighty Media, Inc.
Editor: Katherine Chu
Cover Photographs: Dennis W Donohue/Shutterstock Images (alligator); Sara Louise Singer/ Shutterstock Images (swamp)
Interior Photographs: ad-e-motion/iStockphoto, p. 46; Alan Diaz/AP Images, pp. 42–43; ChiccoDodiFC/ iStockphoto, pp. 38–39; Daniel Carlson/Shutterstock Images, pp. 22–23; Deborah Ferrin/ Shutterstock Images, pp. 34–35; Delmas Lehman/Shutterstock Images, pp. 6–7; Dennis W Donohue/ Shutterstock Images, p. 1 (alligator); Denton Rumsey/Shutterstock Images, pp. 20–21; Haxona/ Shutterstock Images, pp. 18–19; huang jenhung/Shutterstock Images, p. 15; itman__47/Shutterstock Images, pp. 16–17; jaimie tuchman/Shutterstock Images, pp. 14–15; knape/iStockphoto, p. 44; Kyle M. Cooper/Shutterstock Images, pp. 32–33; lazyllama/Shutterstock Images, pp. 8–9; Mariusz S. Jurgielewicz/Shutterstock Images, pp. 4–5; Mark Kostich/iStockphoto, pp. 36–37; Mason Brock/ Wikimedia Commons, p. 33; Odua Images/Shutterstock Images, pp. 12–13; Panchuali/Shutterstock Images, pp. 30–31; PrimeSource Media/Shutterstock Images, pp. 10–11; Raul Baena/Shutterstock Images, pp. 24–25; Red Huber/AP Images, pp. 40–41; Sara Louise Singer/Shutterstock Images, p. 1 (swamp); soleg/iStockphoto, pp. 26–27; Vastram/Shutterstock Images, pp. 28–29
Design Elements: Nik Merkulov/Shutterstock Images (grunge background); queezz/Shutterstock Images (brush texture)

Library of Congress Control Number: 2023939331

Publisher's Cataloging-in-Publication Data
Names: Rusick, Jessica, author.
Title: Surviving the swamp / by Jessica Rusick.
Description: Minneapolis, Minnesota : Abdo Publishing, 2024 | Series: Xtreme survival | Includes online resources and index.
Identifiers: ISBN 9781098291860 (lib. bdg.) | ISBN 9781098278762 (ebook)
Subjects: LCSH: Survival--Juvenile literature. | Survival skills--Juvenile literature. | Swamps--Juvenile literature. | Wetlands--Juvenile literature. | Outdoor life--Juvenile literature. | Wilderness survival--Juvenile literature.
Classification: DDC 613.69--dc23

TABLE OF CONTENTS

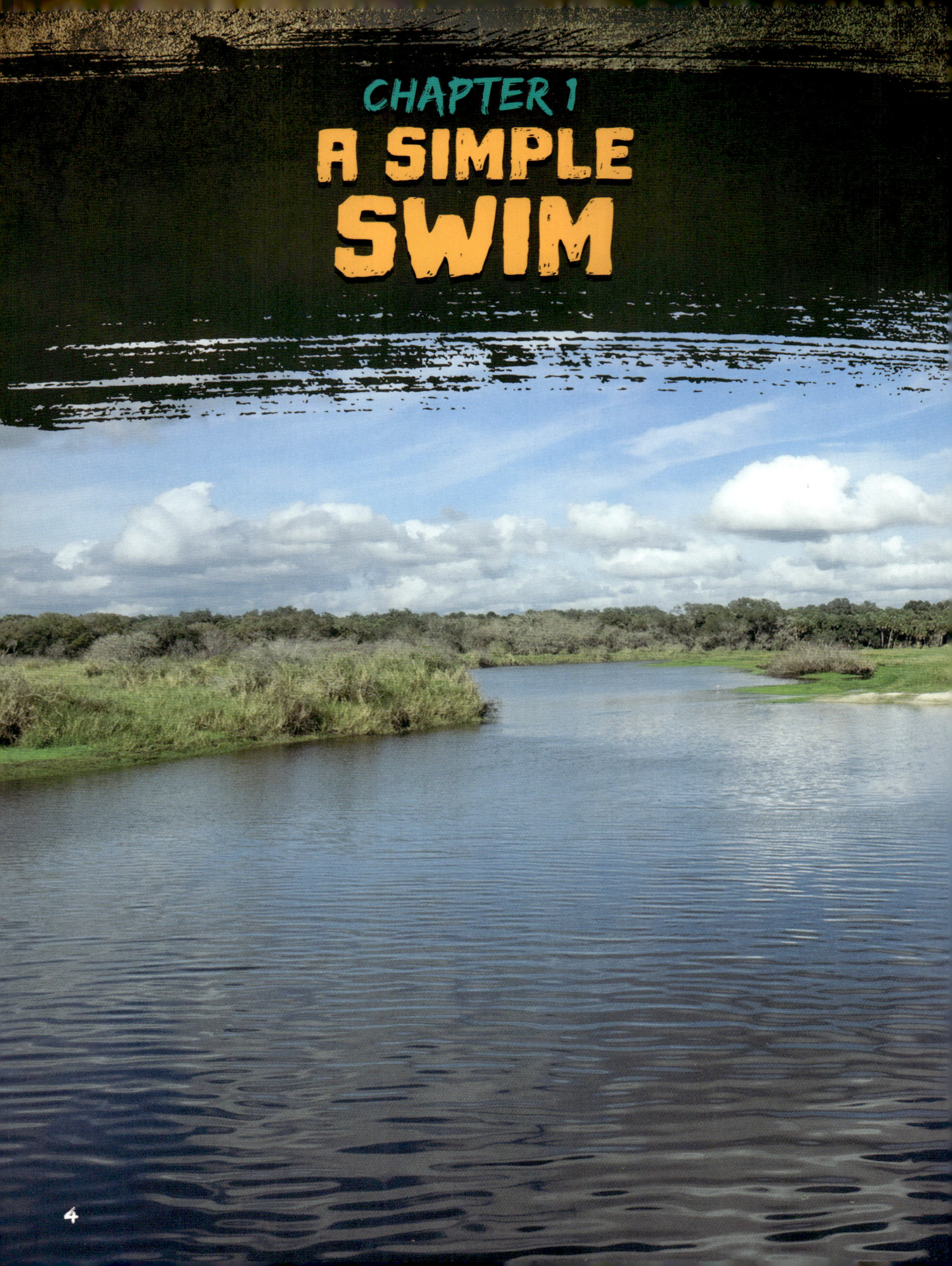

CHAPTER 1
A SIMPLE
SWIM

In 2022, Eric Merda parked his van to explore a Florida swamp on foot. After several hours, the exhausted 43-year-old spotted his van about 1 mile (1.6 km) away on the other side of the swamp. To save time, Merda decided to swim to his van.

Merda had extra time after work and decided to drive down an unknown dirt road, leading him to the swamp.

Merda began swimming across the swamp. But he soon felt something watching him. Suddenly, an alligator bit Merda's right arm, yanking him underwater. It tore off his right arm and swam away. Merda made it back to shore, but he didn't know where he was. He was seriously injured and **stranded** in the swamp. How would he survive?

After Merda's rescue, a trapper found and removed two alligators from the swamp. One was six feet (1.8 m) long and the other was nine feet (2.7 m) long.

The swamp is a **dangerous** place to become **stranded**. Dense undergrowth and mud make it difficult to **navigate**. Toxic plants, hungry alligators, and **venomous** snakes are constant **threats**. But Merda and many others have overcome these dangers and more, thanks to their survival skills.

SURVIVOR SPOTLIGHT

Merda sometimes saw an alligator while walking through the swamp. "It would pop up out of the water, then disappear and reappear," he says. "It was pretty terrifying."

Luckily, Merda's wound was not bleeding. Using his left arm, he pulled himself up against a tree to wave and shout at passing planes. But they did not notice him. Since the undergrowth was thick and thorny, Merda walked near the edge of the swamp. Over the next three days, he ate flowers and drank swamp water to survive. On the fourth day, he saw an empty bottle on the ground. This meant people were nearby! Soon after, Merda spotted a man, shouted for help, and was rescued.

CHAPTER 2

FINDING WATER

Finding water is key to swamp survival. Without water, humans become dehydrated, causing confusion and fainting. Since most swamps are hot and humid, people sweat more. This causes the body to lose water faster, speeding up dehydration.

Humans can survive for about three days without water. Dehydration can lead to hallucinations, seizures, kidney failure, and death.

There are several sources of drinking water in a swamp. One is rainwater. Catch rainwater in bottles or large leaves. Vines also contain water. Cut through the vine in a high place and then a low place, and water will drain out. Do not cut the low end first, or water will be drawn back up the vine. Check that the water is not sticky, milky, or bitter. This means it's unsafe to drink.

If you have two
containers, you
can make your
own water filter.
Layer pebbles,
sand, charcoal,
and cloth to
filter water from
one container to
the other.

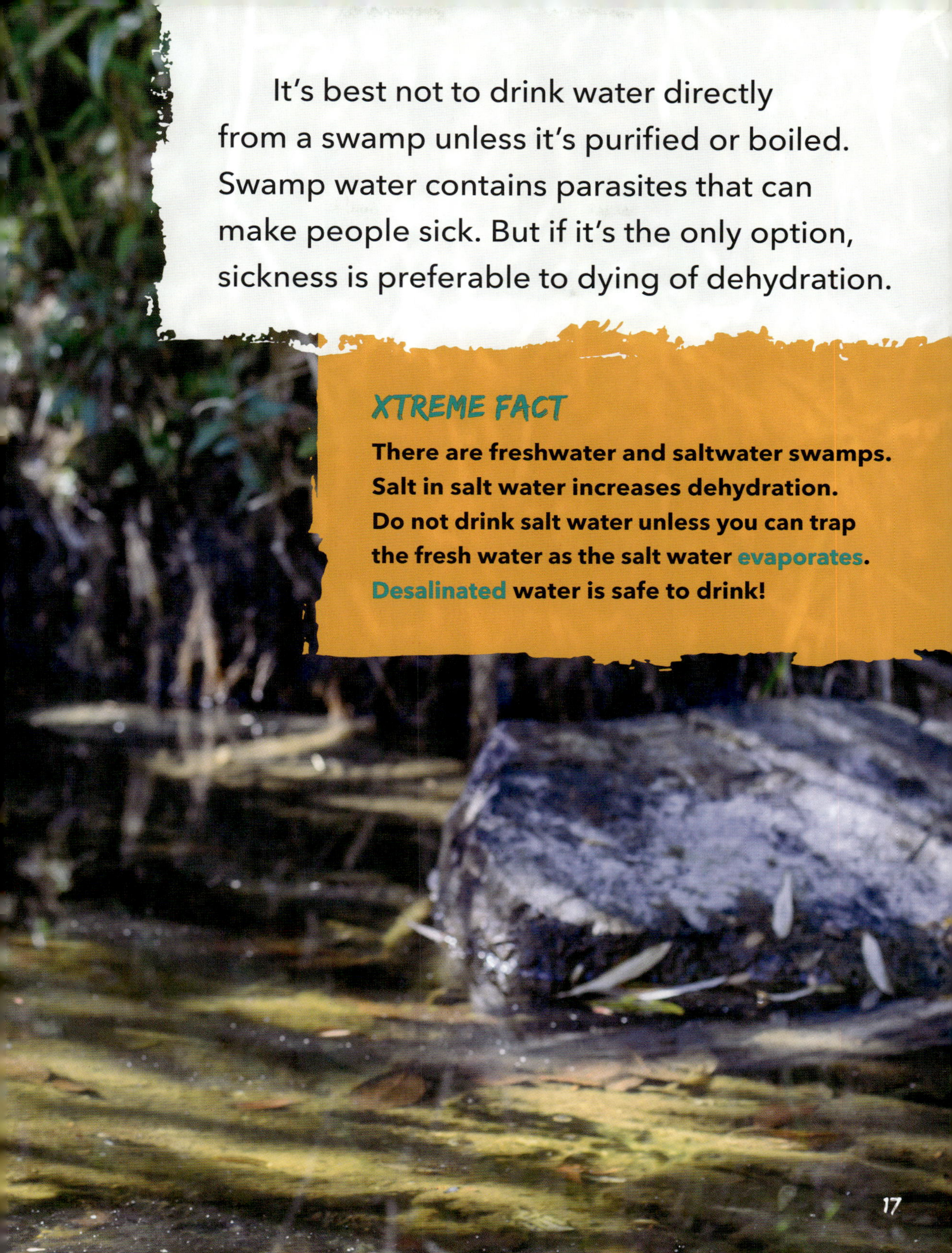

It's best not to drink water directly from a swamp unless it's purified or boiled. Swamp water contains parasites that can make people sick. But if it's the only option, sickness is preferable to dying of dehydration.

XTREME FACT

There are freshwater and saltwater swamps. Salt in salt water increases dehydration. Do not drink salt water unless you can trap the fresh water as the salt water evaporates. Desalinated water is safe to drink!

CHAPTER 3
STAYING DRY

Prolonged exposure to wetness can make hypothermia worse. In wet conditions, it's possible to die of hypothermia even when it is 70 degrees Fahrenheit (21°C)!

Staying dry is challenging in a swamp. But it's necessary to avoid hypothermia. Hypothermia happens when the body's temperature drops, eventually causing the heart to stop beating. It can set in even if it's not very cold.

A shelter can help keep you dry. In 2010, Nadia Bloom got lost in the Florida swamp near her house. The 11-year-old found natural shelters to sleep in at

night, including under a bush and on a tree stump.
She was rescued after four days.

Bloom also slept in a hollow log. This helped keep her safe and dry.

In 2013, the Schreck family went missing in a Florida swamp. Scott, Carrie, and their three sons were **stranded** after their airboat got stuck. Scott built a lean-to shelter to protect them overnight. Make a lean-to by leaning a large branch against a tree. Lean smaller branches against the big branch and cover them with leaves or moss.

Spanish moss is a plant found in some swamps. It hangs off trees in large masses and can be used to pad and insulate shelters!

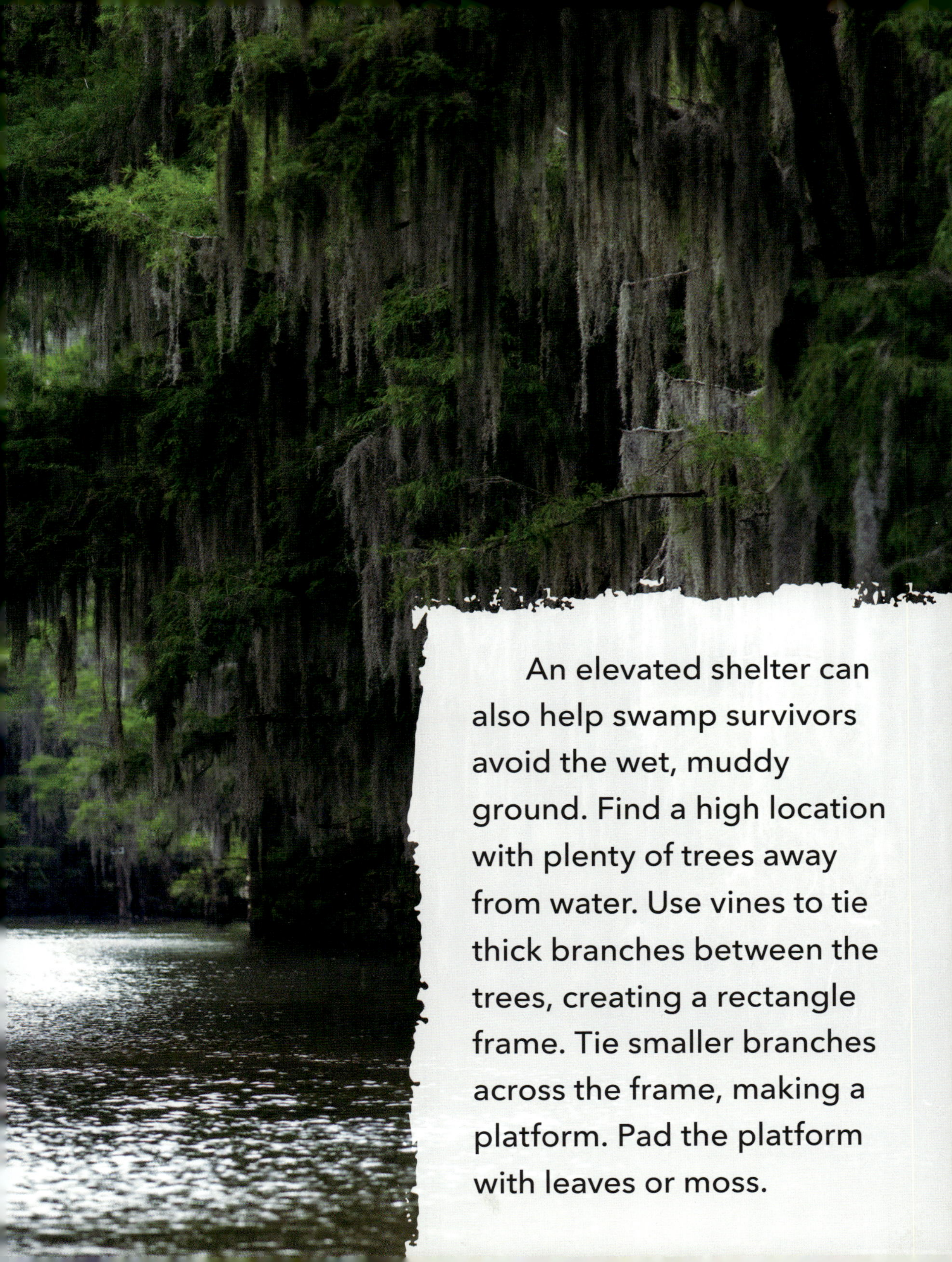

An elevated shelter can also help swamp survivors avoid the wet, muddy ground. Find a high location with plenty of trees away from water. Use vines to tie thick branches between the trees, creating a rectangle frame. Tie smaller branches across the frame, making a platform. Pad the platform with leaves or moss.

Clear the ground and make a fire ring out of sand and rocks. This will keep the fire from spreading and damaging the surrounding area.

Building a fire is another way to stay warm. This can be difficult in a swamp, but it's not impossible. Cut open wet logs and check for dry centers. Build the fire on the dry log pieces. Cattail heads have dry, fluffy seeds that can be used to start the fire!

XTREME FACT

Prolonged wetness can cause fungal infections, such as foot infections. To avoid them, take off your shoes and socks at night to help your feet dry.

CHAPTER 4

FINDING FOOD

Finding food is not as important as finding water and shelter. But there are several swamp animals that are safe to eat. Sharpen a stick and use it to spear turtles, fish, and crawfish.

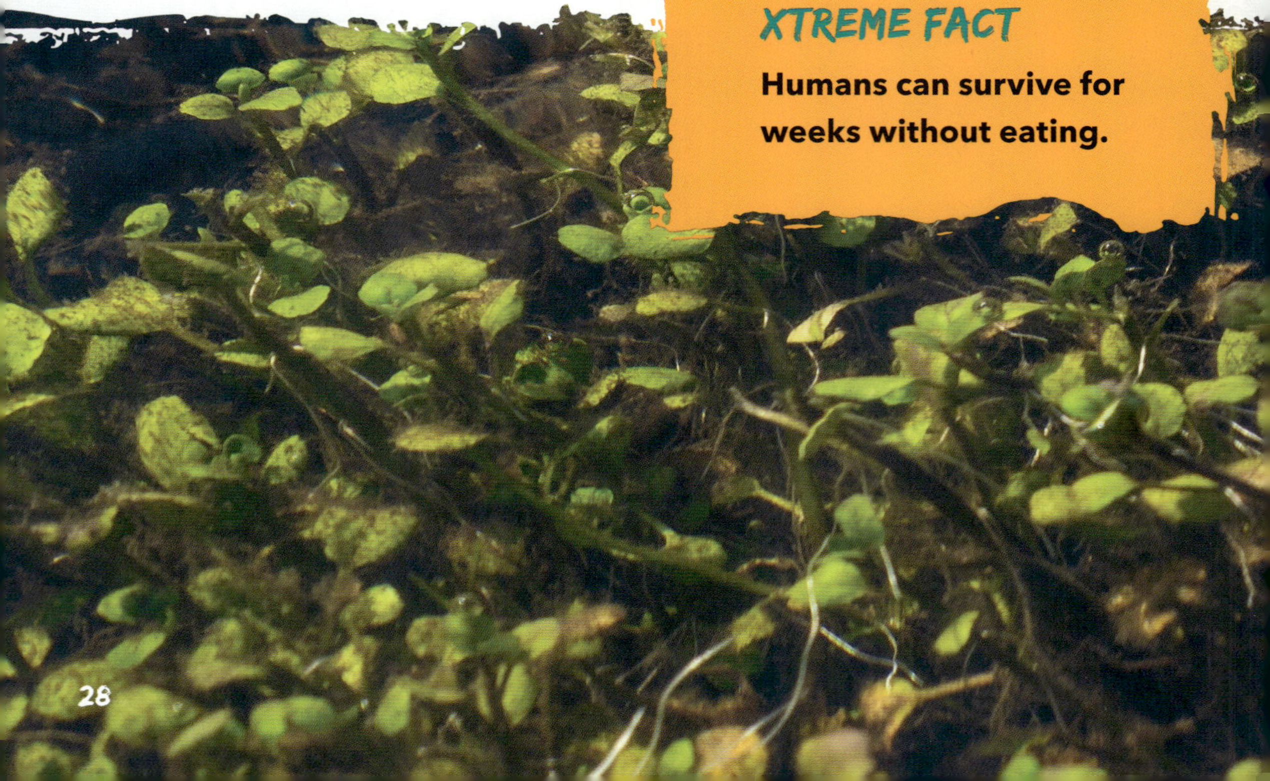

XTREME FACT

Humans can survive for weeks without eating.

Frogs can be caught
by hand and roasted
over a fire.

There are also several edible plants in a swamp, including wild blueberries and cattails. Every part of a cattail is safe to eat. The stems can be eaten raw. Boil the shoots, flowers, and roots before eating.

Swamps also have **dangerous** plants.
Poison ivy and sumac release oils that cause
itchy rashes. Poisonwood trees have black sap
that also causes an itchy rash. And sawgrass
has sharp ridges that can cut through skin!

Poisonwood trees have teardrop-shaped leaves with yellow outlines.

CHAPTER 5

SWAMP DANGERS

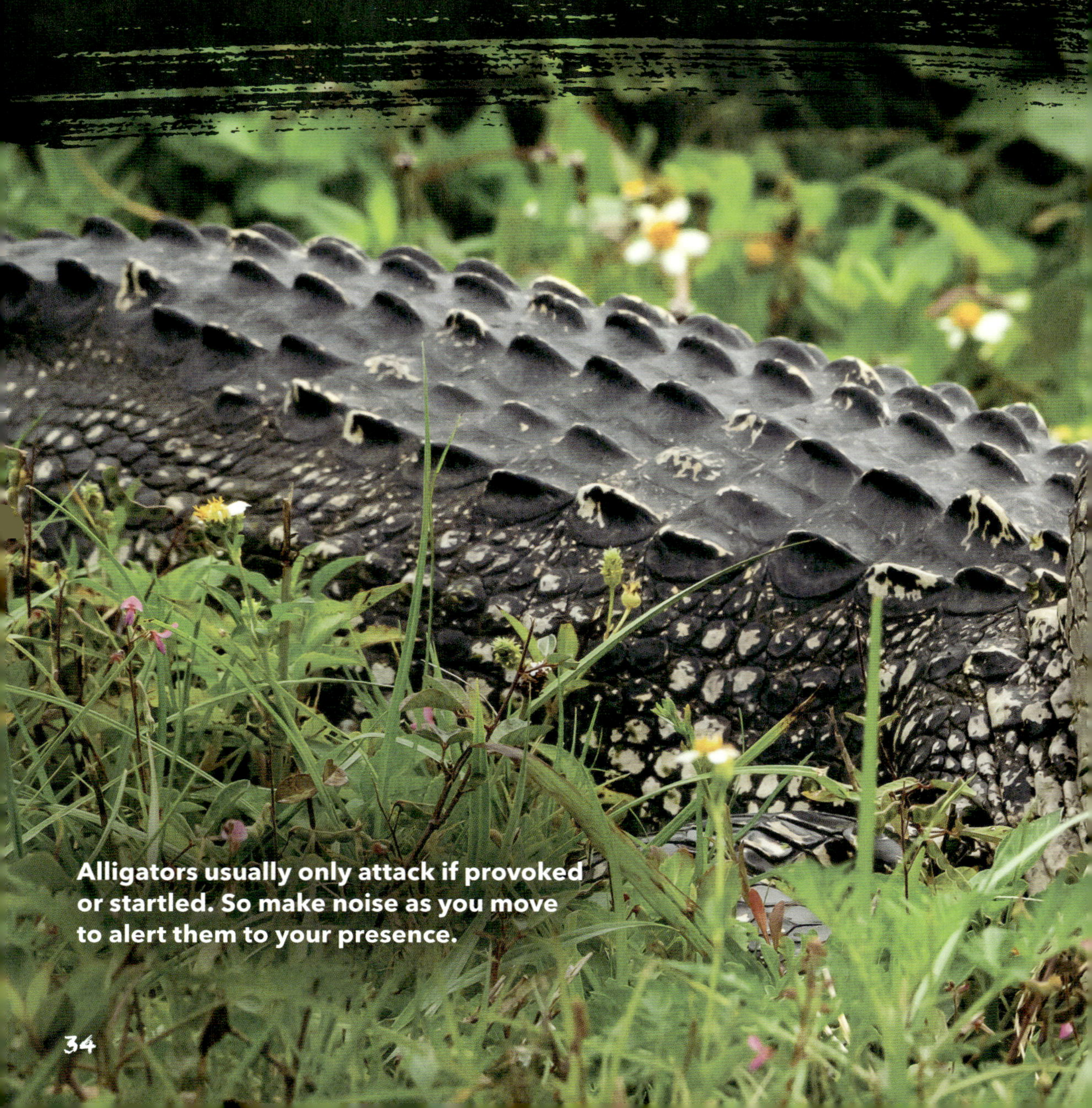

Alligators usually only attack if provoked or startled. So make noise as you move to alert them to your presence.

Alligators are a common swamp danger. They often hide at the edges of water and are most active during spring and summer nights. It's important to avoid the water during these times. The best way to survive an alligator attack is to fight back. Hit the alligator's eyes or jaw, and try to make noise to scare it.

Cottonmouth snakes
are also known as
water moccasins.

Cottonmouth snakes are another swamp danger. These **venomous** snakes live in water and on land. Though cottonmouths usually only bite if they are **disturbed**, their bites are deadly. It's important to be aware of your surroundings. Stay calm and still if you are bitten. This will help slow the spread of venom through the body.

Swamps can be difficult to **navigate**. To avoid injury, walk slowly. Use a stick to probe areas for deep mud. If you fall into deep mud, struggling will only make you sink. Instead, lay the stick horizontally in front of you. Lean over the stick and wriggle your chest onto it. Keep wriggling and pushing on the stick to get free.

Once you free yourself from
deep mud, crawl to solid ground.
If you try to stand right away,
you may get stuck again.

Search teams started looking for Bloom after her mother reported her missing. They left signs encouraging her to stay in place so they could locate her more quickly.

Being **stranded** in a swamp is scary. But survival situations require calm and clear thinking. Bloom prayed, sang songs, and listened to croaking frogs to stay calm. She also stayed in one area. If she had moved around, it would have been more difficult for rescuers to find her.

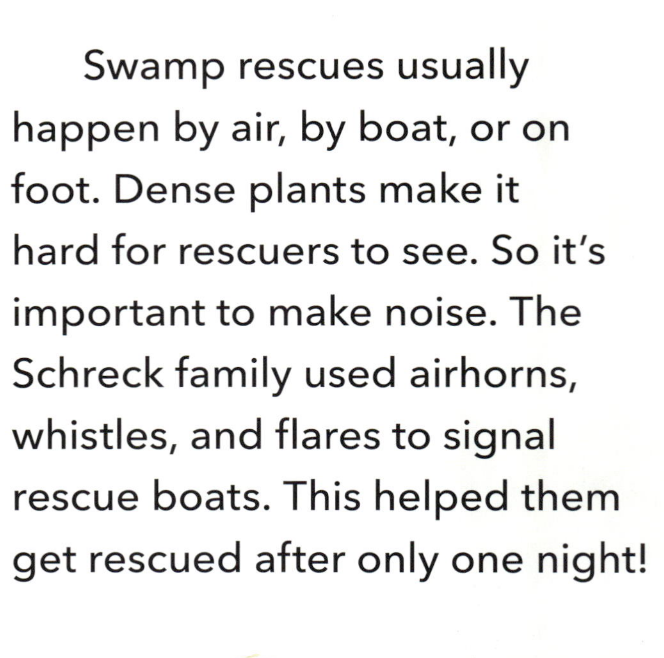

Swamp rescues usually happen by air, by boat, or on foot. Dense plants make it hard for rescuers to see. So it's important to make noise. The Schreck family used airhorns, whistles, and flares to signal rescue boats. This helped them get rescued after only one night!

Scott Schreck (*top right*) and his family after being found. Rescuers used helicopters, airplanes, and airboats when looking for the Schreck family.

CHAPTER 7

SURVIVING THE SWAMP

Being **stranded** in the swamp means having little **access** to fresh water, shelter, and dry conditions. Alligators and thick mud are constant dangers. But with top-notch survival skills and a bit of luck, some people have overcome these challenges to survive. Think back on the swamp survival stories you read and skills you learned. Could you survive in the swamp?

When visiting a swamp, it's best to stay on visible and marked trails. If you decide to go off-trail, bring a walking stick to avoid slips or falls.

XTREME CHALLENGE

1) How did Eric Merda know he was near people in the swamp?

2) How do you build a fire in wet conditions?

3) If you were stranded in a swamp, would you rather have a bottle for water or a tarp for shelter? Why?

4) Why does sweating speed up dehydration?

5) How would you stay calm if you were stranded in a swamp?

GLOSSARY

access—the opportunity to gain or use something.

dangerous—able or likely to cause hurt or harm.

desalinate—to remove the salt.

disturb—to bother or interrupt.

evaporate—to change from a liquid into a gas or vapor.

fungal—caused by fungus. Fungus is an organism, such as mold or mildew, that grows on organic matter.

infection—a disease caused by the presence of bacteria or other germs.

navigate—to find the way from place to place.

stranded—lacking the means to leave a place.

threat—something that could be harmful.

venomous—producing poison. Venom is a poisonous substance produced by some animals and usually injected into victims through a bite or sting.

ONLINE RESOURCES

To learn more about swamp survival, please visit **abdobooklinks.com** or scan this QR code. These links are routinely monitored and updated to provide the most current information available.

INDEX